Art Department

ceramics by COIL

by Joan & Anthony PRIOLO

LITTLE CRAFT BOOK SERIES

STERLING PUBLISHING CO., INC. NEW YORK
Oak Tree Press Co., Ltd. London & Sydney

Little Craft Book Series

Aluminum and Copper Tooling
Animating Films without a Camera
Appliqué and Reverse Appliqué
Balsa Wood Modelling
Bargello Stitchery
Beads Plus Macramé
Beauty Recipes from Natural Foods
Big-Knot Macramé
Candle-Making
Cellophane Creations
Ceramics by Coil
Ceramics by Slab
Corn-Husk Crafts
Corrugated Carton Crafting
Costumes from Crepe Paper
Crafting with Nature's Materials
Creating from Remnants
Creating Silver Jewelry with Beads
Creating with Beads
Creating with Burlap
Creating with Flexible Foam
Creating with Sheet Plastic
Creative Lace-Making with Thread and Yarn
Cross Stitchery
Curling, Coiling and Quilling
Decoupage—Simple and Sophisticated
Embossing of Metal (Repoussage)
Enamel without Heat
Felt Crafting
Finger Weaving: Indian Braiding
Flower Pressing
Folding Table Napkins
Greeting Cards You Can Make
Hooked and Knotted Rugs
Horseshoe-Nail Crafting
How to Add Designer Touches to Your Wardrobe
Ideas for Collage
Junk Sculpture
Lacquer and Crackle
Leathercrafting
Macramé
Make Your Own Elegant Jewelry
Making Paper Flowers
Making Picture Frames
Making Shell Flowers
Masks
Metal and Wire Sculpture
Model Boat Building
Monster Masks
Nail Sculpture
Needlepoint Simplified
Net-Making and Knotting
Off-Loom Weaving
Organic Jewelry You Can Make
Patchwork and Other Quilting
Pictures without a Camera
Pin Pictures with Wire & Thread
Potato Printing
Puppet-Making
Repoussage
Scissorscraft
Scrimshaw
Sculpturing with Wax
Sewing without a Pattern
Starting with Stained Glass
String Designs
String Things You Can Create
Tissue Paper Creations
Tole Painting
Trapunto: Decorative Quilting
Whittling and Wood Carving

Acknowledgments

The authors and publishers would like to thank artist Oscar Bucher of Santa Barbara, California, for permission to show his work in this book. Thanks also to Wayne McCall of Santa Barbara, who did the photography for the book.

Second Printing, 1977

Two Park Avenue, New York, N.Y. 10016
Distributed in Australia and New Zealand by Oak Tree Press Co., Ltd.,
P.O. Box J34, Brickfield Hill, Sydney 2000, N.S.W.
Distributed in the United Kingdom and elsewhere in the British Commonwealth
by Ward Lock Ltd., 116 Baker Street, London W 1
Manufactured in the United States of America

Library of Congress Catalog Card No.: 76–1178
Sterling ISBN 0-8069-5344–6 Trade Oak Tree 7061-2103–1
5345–4 Library

CONTENTS

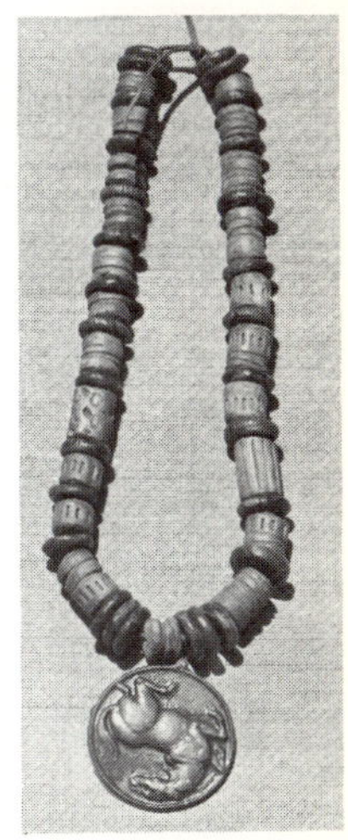

Before You Begin

The coil method is one of the oldest ways of making ceramics without a potter's wheel. Ceramics made with coils have a unique charm which cannot be achieved in any other way.

Illus. 1. Here are some of the coil projects you will make.

Although you can make coil ceramics in the traditional manner (see Make a Free-Form Coil Pot, page 9), you will learn here on these pages another and easier way using simple, porous moulds. You can purchase these moulds as paper (*unsized* and *unwaxed*) paint containers, paper plates, wooden planters (also *unsized* and *unpainted*) and cardboard boxes—all everyday household items. We will also show you how to make your own plaster moulds.

Once you have learned how to roll out or cut coils (did you know that you can make a simple tool that will cut coils almost 50 times faster than rolling them by hand?), don't hesitate to combine them with clay slabs and "buttons" for interesting design variations.

Keep in mind that the charm of coil work lies in its irregularity, so do not try to place perfect coils in a perfect pattern. Keep it loose and let go. Your coil projects can include lamp bases, cookie jars, planters, tiles, mirror frames, hanging lamps, jewelry and buttons, to name a few.

You can use almost any type of modelling clay for coil work. However, the clay should contain grog (pre-fired particles of clay) to help prevent the clay from cracking when bisque-fired.

To make your coil projects even easier and more fun, we suggest the use of a simple and inexpensive method of glazing which does not require a kiln firing and which can be applied to bisque-fired ware.

We hope that the ideas offered in this book will inspire you to make working with coils an exciting and new adventure!

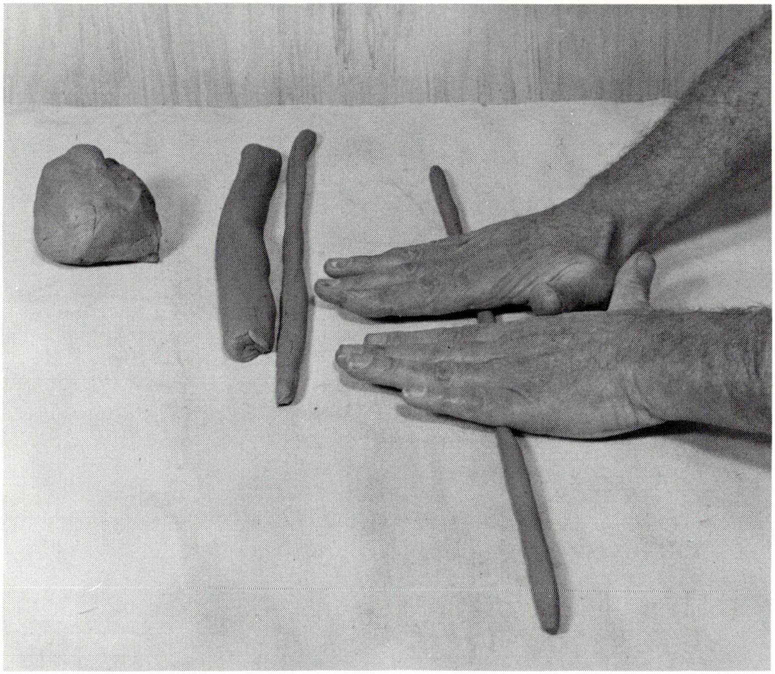

Methods and Materials

Illus. 2. From a ball of clay, first make a thick roll, and then cut several pieces from it to roll your coils.

Rolling Out Clay Coils

Be sure that the clay you use is fairly soft (wet). To soften clay, simply knead water into it.

First, cut off a desired amount of clay, and pat it into an even ball with your hands. Place the ball on newspaper, the nonslick side of oil cloth, or an unsized piece of canvas. Then roll the clay ball, using the palms of your hands, into a thick (1-inch diameter) roll. Cut this thick roll into lengths of 3 to 4 inches, and roll these pieces into coils of the desired thickness, but no longer than 15 to 20 inches. (See Illus. 2.)

When rolling the coils it is important that you use a *back-and-forth* movement, rather than too much down pressure which tends to flatten the coils. This will take practice. As you become more proficient, you can use longer pieces to make longer coils. *Complete uniformity of the coil is not necessary*—you will find that variations of thickness often add to the interest of a finished project.

We suggest starting out with a very small object in order to get an idea of the number of coils you will need for different kinds of projects, since it is best to have all your coils made (plus extras for insurance) before starting. Place your prepared coils on a piece of plastic and cover with another piece of plastic to keep them soft and wet.

Cutting Coils with a Hand-Made Tool

If you are planning to make quantities of coil objects, or simply want an easier and quicker way to make coils, you can make a coil-cutting tool.

The easiest way to make a tool is to buy a modelling tool with a brass wire end. With a small pair of pliers, open the brass end into a round shape. Slip a piece of copper, brass, or aluminum tubing (available at hardware dealers) inside the brass wire and, with pliers, mould the wire round the tubing to form a perfect circle (Illus. 3). With the tubing still in position, bind the ends of the wire closed with electrical or adhesive tape or thin copper wire—or both. When the tubing is removed, you will have a perfect circle. File the wire circle flat on the inside and outside for smooth cutting.

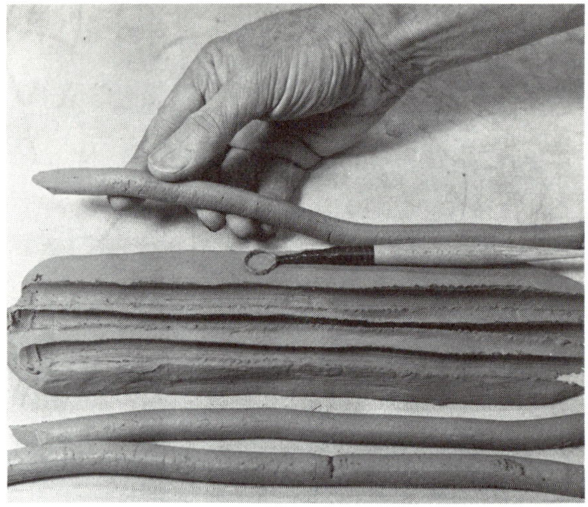

Illus. 4. Cutting coils with the cutting tool.

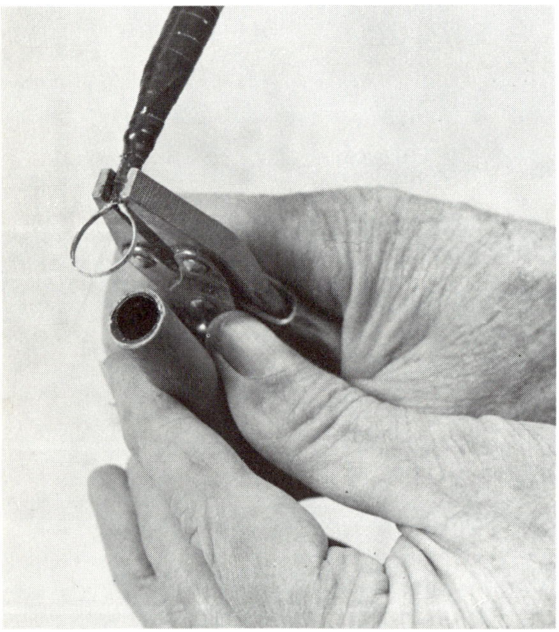

Illus. 3. Making a coil-cutting tool.

You can also make a cutting tool with 12- to 14-gauge brass wire by embedding the two ends of the wire in a wooden handle. Follow the above procedure of forming, taping, and filing.

To cut coils with your hand-made tool, roll (see Illus. 39) out a thick (approximately 1-inch) slab of clay the length you want for the coils. The width of the slab should be about 4 to 5 inches. Draw the wire circle of the cutting tool through the length of the clay slab to make the coil (Illus. 4). Repeat the cutting until you have cut as many coils as the width of the slab permits. With this method you can make coils almost 50 times faster than by hand!

Kneading

Once you have your clay coils in place, you will have to bind them together by kneading on the *inside* of your object (Illus. 5).

Knead with your forefinger or thumb, using an up-and-down motion. Be careful not to exert too much outward pressure if you are using a mould, or the outside coil design will be flattened. If you are kneading a free-form pot, too much outward pressure may cause the walls to collapse.

The coils must be *soft* and *damp* or the kneading will be difficult to do—also, the pot might crack when dry. To make the coil object stronger and to help the kneading, you may want to use extra clay to "plaster" the inside walls. This "plastering" clay should have the consistency of soft butter (Illus. 5).

A Reinforcing Coil

When you have finished kneading the inside of your project, it is a good idea to add an extra coil round the inside top of the object (Illus. 6). This extra coil will add strength and prevent cracking. Knead the coil (or coils if your project is large) and blend it with the inside surface. Be sure to add a reinforcing coil to all large and thin-walled objects.

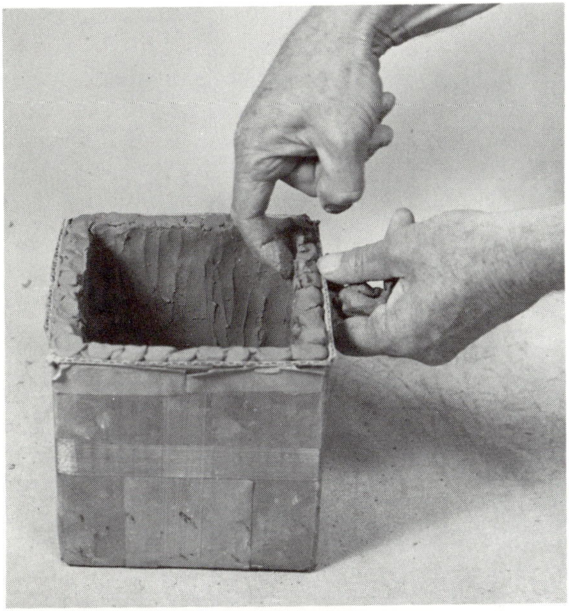

Illus. 5. Knead the coils together with your forefinger or thumb, using an up-and-down motion.

Illus. 6. Add a reinforcing coil to the inside top of your coil pot.

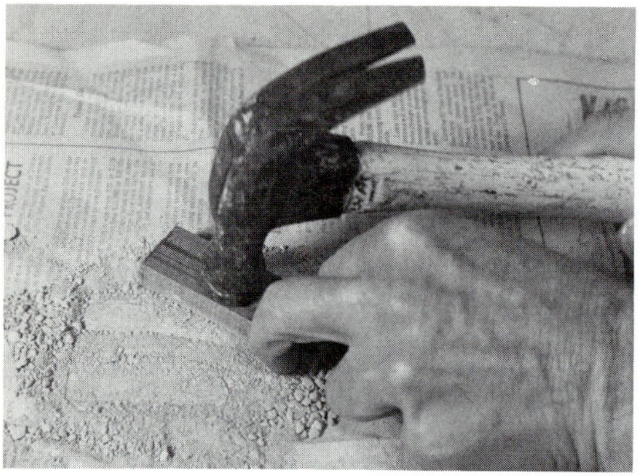

Illus. 7. You will need powdered clay to use as a parting agent for paper or wooden moulds.

Powdered Clay

Powdered clay is used as a parting agent for paper or wooden moulds. Dry a small piece of clay by letting it stand in the sun until bone dry, or by baking it dry in a kitchen oven set at between 300° and 400°F. (149° and 205°C.) for about half an hour. Then place the dry clay on a sheet of newspaper and crush it with a hammer and a small block of wood (Illus. 7) until it is powder-fine.

Slip

You will use slip, or liquid clay, to attach one piece of clay to another. Make slip by letting a small quantity of clay dry hard; then place it in a paper bag and hammer the bag lightly until the clay is broken into small bits. Place the crushed clay in a can and slowly add water, stirring until it has a soft, buttery consistency.

Before cementing with slip "score," or roughen, the edges of the clay with a tool to assure a strong bond.

Storing Clay

If you spray your clay with water at intervals and keep it covered with a plastic bag, you can keep a clay project workable for anywhere from a few days to a few weeks. Either wrap the piece in a wet cloth or place a damp sponge in the plastic bag (Illus. 8).

Bisque Firing

You will need to bisque-fire your finished coil projects in a ceramic kiln. Then you will be ready to glaze them with a ceramic glaze or with a non-firing glaze. For details on firing and glazing, see page 37.

Illus. 8. Keep your clay workable in a "damp closet," consisting of plastic and a wet sponge.

Illus. 9. With a clay coil, make a small circle for the base of your free-form pot.

Make a Free-Form Coil Pot

Roll out or cut enough coils for a small pot. Cover a small turntable (Lazy Susan or a piece of wood that can turn without disturbing the work in progress) with construction paper or several sheets of newspaper. Make a base as shown in Illus. 9.

Now, start laying clay coils, log-cabin fashion, one on top of another, to a height of about 2 to 3 inches. Use only enough pressure to make contact. Pat each coil gently onto the other. Too much pressure will collapse the pot, especially as it gets taller.

Before adding more coils, knead the coils together using the forefinger or thumb of one hand while holding the wall in place with the other hand, using a little inward pressure, directly opposite the kneading finger (Illus. 10).

To make a straight-walled pot, place the coils directly on top of each other. To make a rounded

Illus. 10. Knead the first 2 to 3 inches of coils on the inside of the pot.

Illus. 11. Then lay another 2- to 3-inch height of coils and knead the inside again. The entire pot is done this way—laying a 2- to 3-inch height of coils and then kneading.

Illus. 12. To make a rounded free-form pot, place each coil slightly off-center outward.

Illus. 13. You can do some shaping by hand, but do it carefully.

Illus. 14. Add coil handles by pressing them on firmly with your fingers and applying slip. See the finished pot in color Illus. A on page 14.

pot as shown in Illus. 12, place each coil slightly off-center outwardly, but with restraint because the kneading process will automatically push the walls of the pot outward a little.

Although the shaping of the pot is achieved mainly by the placement of the coils, a little shaping can be done by hand by pushing the coils inward or by pushing them outward (Illus. 13). Do this with care or the walls could become too thin and the kneaded coils could come apart. To finish the pot, blend the last coil to form an even top.

Make a Coil Pot Inside a Paper Container

Another very simple way to make a coil object is by using a mould, in this case an *unsized* and *unwaxed* paper paint container. Using a mould eliminates any danger of the walls collapsing while you are working. An advantage to using paper moulds, in addition to the ready availability of paper containers, is that you can draw a design on the outside as a guide to the placement of the clay on the inside.

Illus. 15. First, draw a design on the outside of a paper container and prepare clay coils and small coil "buttons," as shown here. Then line the bottom of the container with a circle of construction paper or newspaper and dust the inside walls with powdered clay (see page 8).

Illus. 16. Now, place your clay coils on the bottom and up the sides of the container, and the clay "buttons" according to the design you drew on the outside. Take care not to press the coils and "buttons" too firmly against the sides of the container or the design will become flattened.

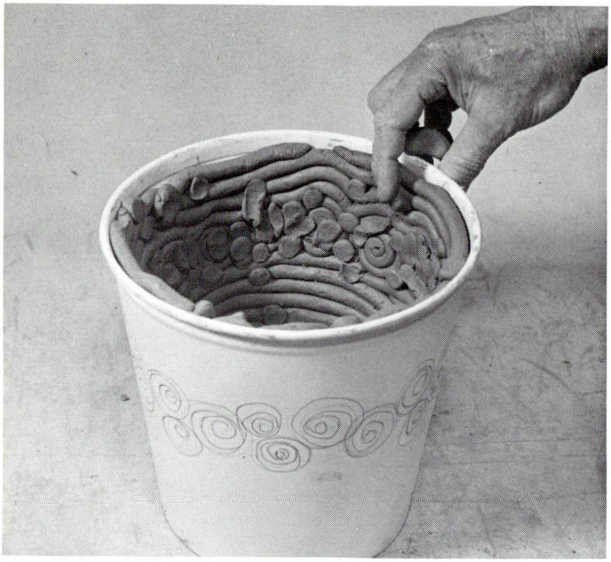

Illus. 17. When working with an irregular design, such as one involving long coils and coil buttons, you will find gaps which you can fill with small wads of clay. Make sure that all of the gaps are filled before proceeding any further.

Illus. 18. After all the gaps have been filled, knead the coils and buttons together with your forefinger, using an up-and-down motion. Again, be careful not to press the clay too firmly against the walls of the container. If any area seems too thin, use more soft clay to "plaster" the inside walls. This will make the walls thicker and stronger. When the inside has been thoroughly kneaded and "plastered," add a reinforcing coil to the top.

Illus. 19. As soon as the clay is leather-hard dry (overnight), place a piece of cardboard on the top of the container and turn the whole thing upside-down. If the paper container does not lift off easily, the clay has not shrunk enough and should be left in the container longer.

Illus. 20. After removing the container and the paper circle from the bottom of the pot, cut a hole in the bottom for drainage. See the finished pot in color Illus. D on page 15.

Illus. A. Handles and textured "buttons" of clay lend personality to this free-form coil pot. (See page 9.)

Illus. B. A large free-form pot by Oscar Bucher. The artistic arrangement of the coils makes this an unusually beautiful pot. (See the close-up on the front cover of this book).

Illus. C. Small wads of clay form the main structure of this container which was made inside a square cardboard box. Circles of clay coils add interest to the over-all design. A single coil was used as the lid handle. (See page 20.)

Illus. D. The irregular arrangement of coils and coil circles provides design interest for this pot which was formed inside a paper container. (See page 11.)

Texture Tools

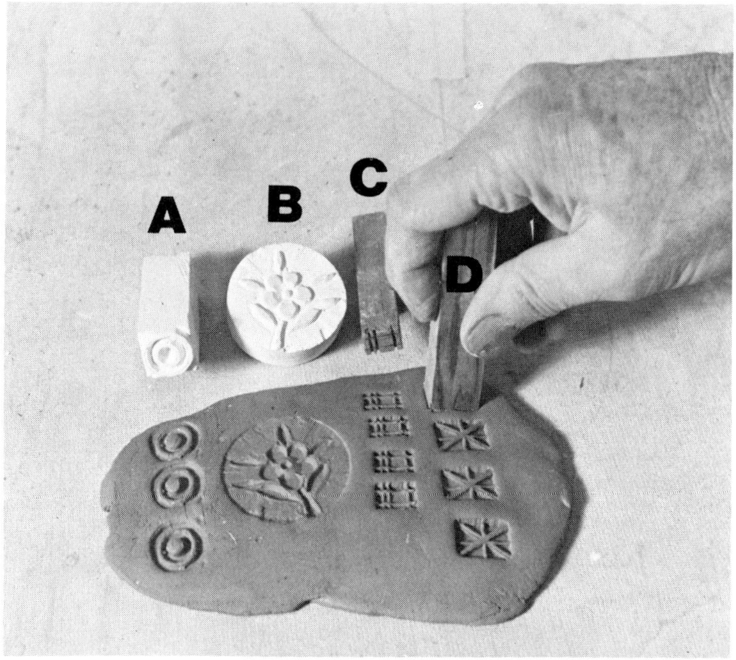

Illus. 21. Some tools for texturing and their resulting designs: A. plaster block with a carved design; B. plaster circle with a carved flower design; C. wooden texture tool with a line design; D. large wooden texture tool with a starlike design.

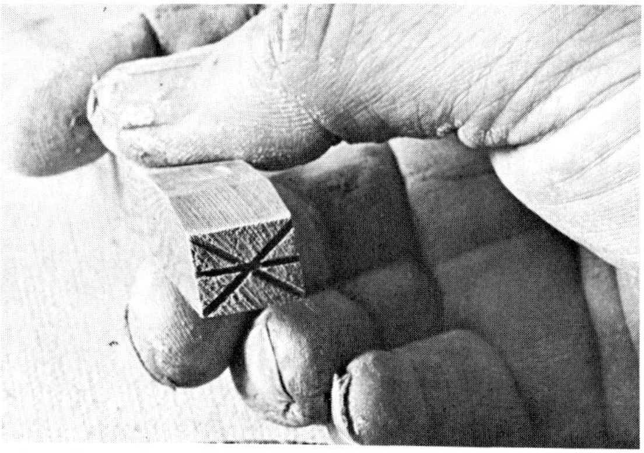

Illus. 22. In order to make a simple wooden texture tool, use a file to groove a design in a small piece of wood.

Illus. 23. Then press the wooden tool into the soft clay to form a repeat design such as this.

Illus. 24. Nature offers many objects that you can use to texture your projects—leaves, twigs, bark, and so on. In addition, ordinary household objects that have texture can be used—nuts and bolts, coins, nails, textured wallpaper, paper clips, brushes, to name just a few.

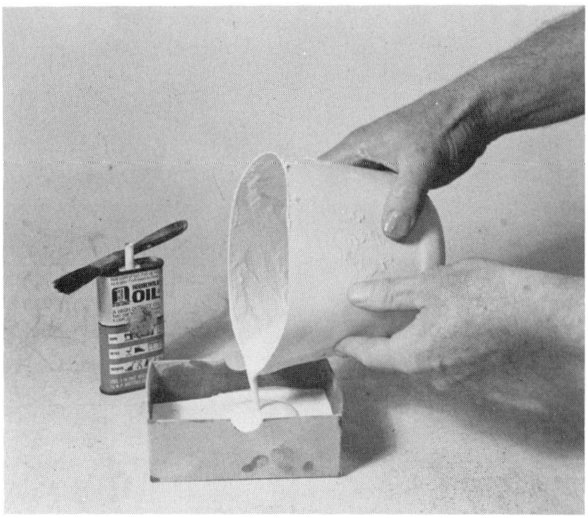

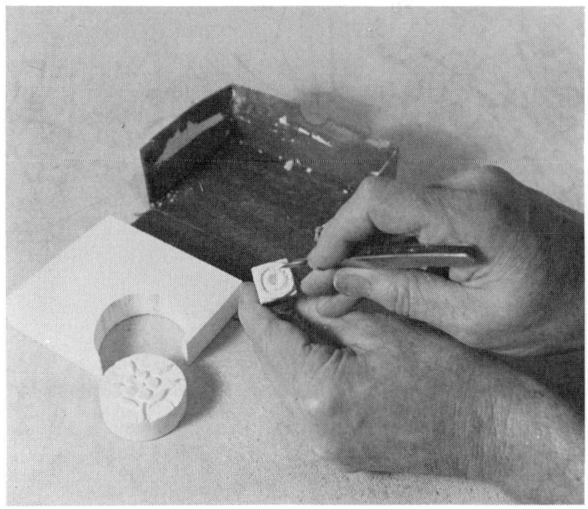

Illus. 25 (left). As you can see in Illus. 21, carved plaster blocks are an especially effective means of stamping special designs into the soft clay. To make a plaster block, mix equal portions of water and plaster of Paris by weight. Stir the mixture for one minute, and then pour it into a cardboard box which you have greased with cooking oil or motor oil, and let the plaster set until dry. Peel off the cardboard box, and you will have a block of plaster which you can cut into smaller blocks or shapes with a saw or knife. Illus. 26 (right). Using a penknife or a carving tool, you can then carve any design you wish into one end of each little block.

Illus. E. This handsome lamp base, made in a paper container, combines coils with slabs that were textured with leaves. (See page 28.)

Illus. F. A tile wall plaque which derives its charm from the loose and random arrangement of textured coils, small slabs, and "buttons" of clay. Tiles of this type would also make a unique table-top, counter-top, or—for the ambitious—even a wall! (See page 39.)

Illus. G. A combination of slabs and random coils makes an unusual planter. The interesting pattern on the slab was obtained by pressing a plant branch into the soft clay. Varied designs can be made by using leaves, flowers, branches, weeds, and other objects from Nature. This pot is the same as the lamp base in Illus. E, put to a different use.

Illus. H. A colorful coaster made with coils and stamped designs is a simple and effective project. (See page 16 for texture tools.)

Illus. I. A simple tile of textured coils, small slabs, and "buttons" makes a decorative and useful trivet. Several tiles could be used to make a handsome tray.

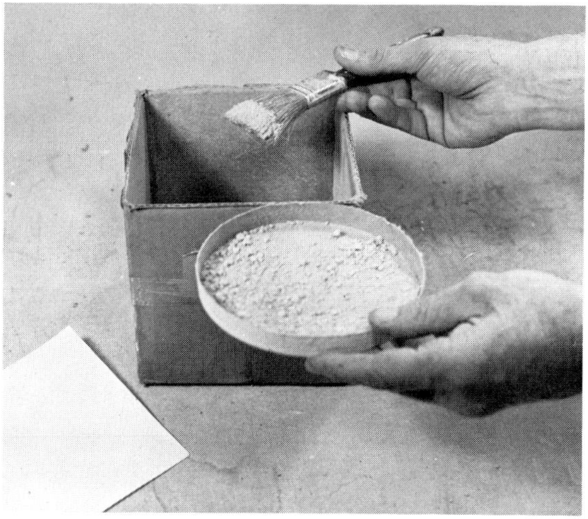

Illus. 27. Dusting the inside of the box with powdered clay.

Illus. 28. Begin building the walls with small wads of clay.

Make a Container and Lid in a Cardboard Box

Dust the inside of a cardboard box with powdered clay, and then line the bottom of the box with a piece of paper (Illus. 27). Prepare a number of circles from clay coils to use for the main design. Now, place a slab of clay in the bottom of the box and start building the clay walls with small wads of clay (Illus. 28). You will use these wads to build the entire box. This is a very simple and effective method to use in combination with coils, slabs, and "buttons."

Place the wads in a fish-scale fashion, that is, with each wad overlapping and making contact

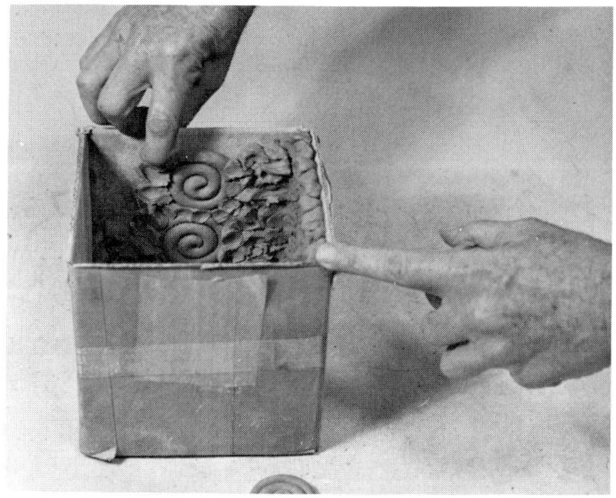

Illus. 29. Place the coil circles in the middle of each wall for interest. Build up the walls to the top of the container and knead the coils and circles together.

Illus. 30. When the clay is leather-hard (overnight), place a piece of cardboard on top of the box, turn the box upside-down and lift it off the clay. Do not force the box. If it does not lift off easily, let the clay dry longer.

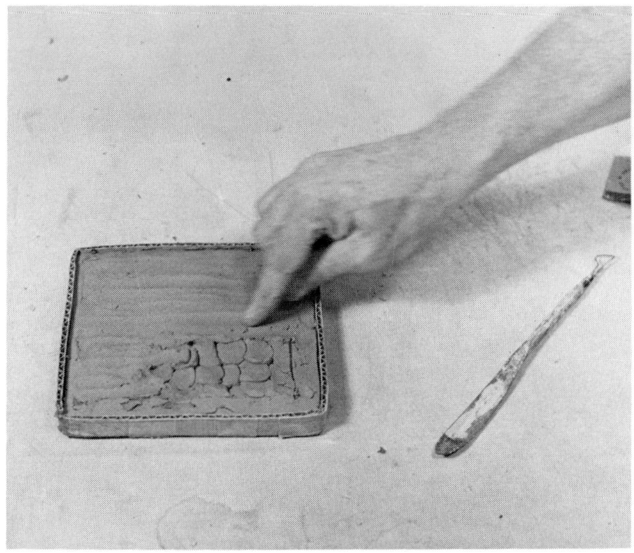

Illus. 31. To make a lid, use the same cardboard box and cut off the bottom, leaving sides of about $\frac{1}{2}$ inch high. Dust the cardboard with powdered clay and cover the area with wads of clay in the same manner as done on the walls. Knead the wads together, taking care not to exert too much down pressure. Then . . . (see Illus. 32)

Illus. J. The necklace was made from short lengths of textured coils and round "buttons." The buttons were cut from a separate coil. (See page 25 and the front cover of this book.)

Illus. K. This coil bowl was formed inside a plaster mould. The deep blue provides a colorful contrast to the rich brown of the outside of the bowl. (See page 36.)

Illus. L. Coils and slabs combine in an irregular pattern to make this handsome planter. A wooden planter was used as a mould. (See page 42.)

Illus. M. Coils and small "buttons" are arranged in a flower design on this small pot. A variety of designs can be achieved with this technique, such as lettering, faces, animals, and so on.

Illus. 32. . . . make four wads of clay to use as flanges which will fit inside the box to prevent the lid from moving. Lightly outline the inside edge of the box on the clay and lay the flanges along this edge, using slip to adhere them.

only at the edges of the preceding wad. Do not apply too much pressure or you will flatten the design.

Illus. 33. When the clay is leather-hard, remove the cardboard. Attach a coil handle to the lid with slip and press it into place with a texture tool. Do not exert too much pressure when attaching the handle or the lid may become distorted.

Illus. 34. Here is the finished container with its lid in place (also see color Illus. C on page 15). This method can be used equally well in making very small or very large objects, depending upon the size of your cardboard container.

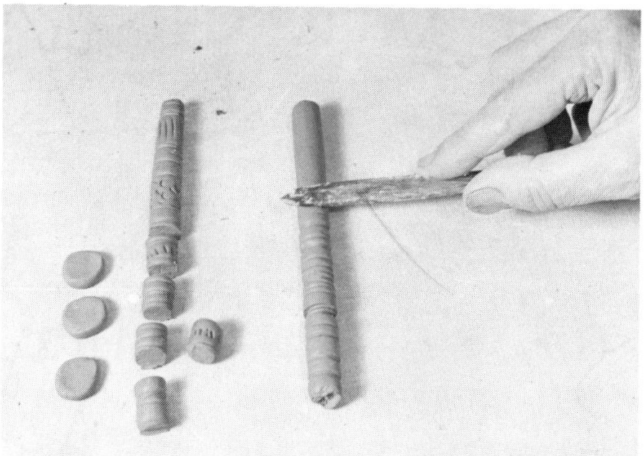

Illus. 35. Texture two clay coils with a modelling tool by placing the tool at intervals on the coil while rolling the coil back and forth. Cut the coils into small lengths, and further texture the pieces by pressing a matchstick and the blunt end of a nail into the clay.

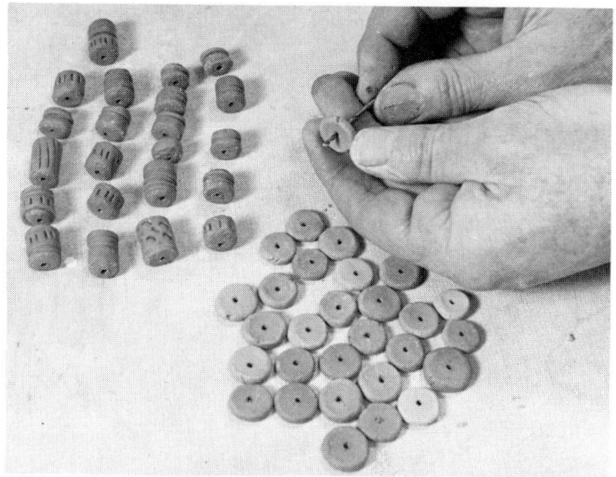

Illus. 36. Then cut small flat "buttons" from another coil. When all the pieces are "soft-leather" dry, drill holes for stringing with a small steel drill. You can use a nail, but it is difficult to push through the clay and might distort the shape if the clay is too soft, or break it if the clay is too hard.

Beads and Buttons

For this kind of project, have your clay a little drier than usual. It must be dry enough so that the design and shape are not distorted by handling. It is a good idea to bisque-fire items intended for wear at a little higher temperature (100° to 200°F., or 38° to 93°C.) than normal for more durability. See the finished beads and buttons in color Illus. J on page 22.

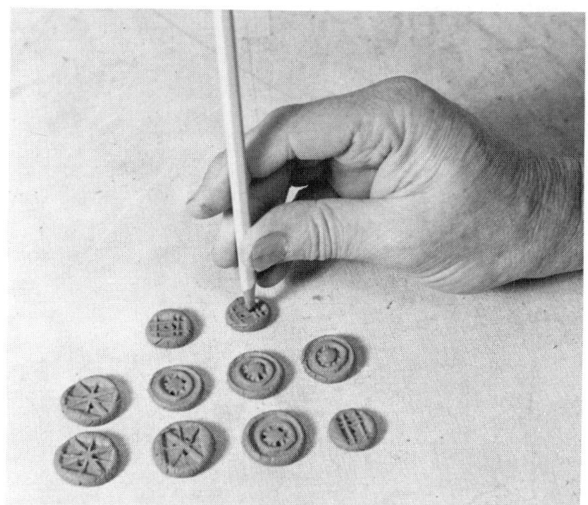

Illus. 37. To make buttons, slice a coil into small sections and press them with your fingers into semi-flat buttons. They can be textured and made even flatter. Punch holes for sewing into the soft clay with a pencil.

Illus. N. A plaster mould was used to form this hanging planter. The filigree effect is achieved by leaving open areas in the coilwork. The use of two colored glazes enhances the coil pattern. Ceramic planters of this kind are especially suited for use with macramé hangers. (See page 36.)

Illus. O. A free-form pot with a smooth finish. Thin slabs of clay with an incised design and stamped "buttons" add interest to a simple pot such as this. (See page 40.)

Illus. P. Open coilwork combined with textured slabs makes an attractive candle holder. A bright blue glaze enhances the filigree effect.

Make a Lamp Base in a Paper Container

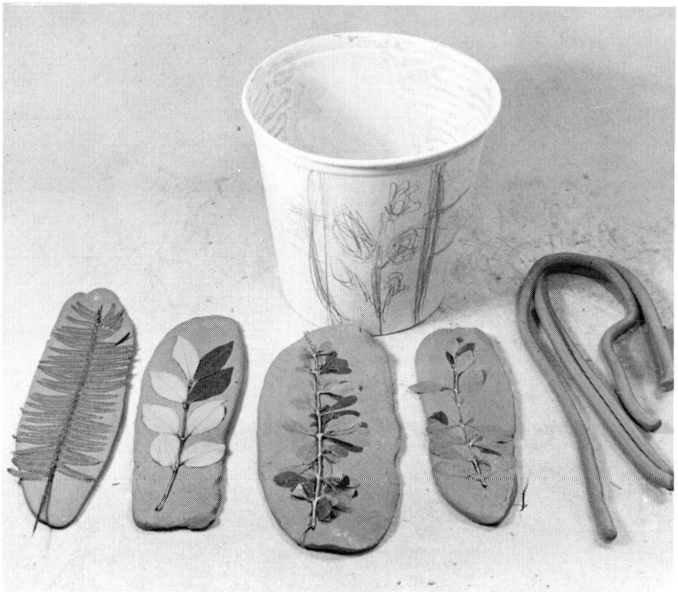

Illus. 38. Dust the inside of an unsized and unwaxed paper container with powdered clay, and place a circle of paper on the bottom. Embed leaves in four slabs of clay the length of the height of the container and draw a rough design on the outside of the container as a design guide. (See Illus. 39 for slab-making.) Then prepare a quantity of clay coils.

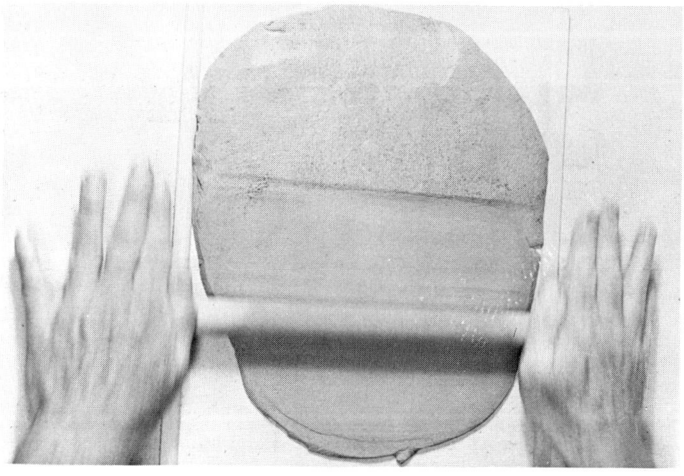

Illus. 39. To make a slab, flatten out a ball of clay with your hands. Then roll it into a slab with a rolling pin or wooden dowel. For an even thickness, place two sticks of wood of the thickness desired for the clay parallel to each other on either side of the slab.

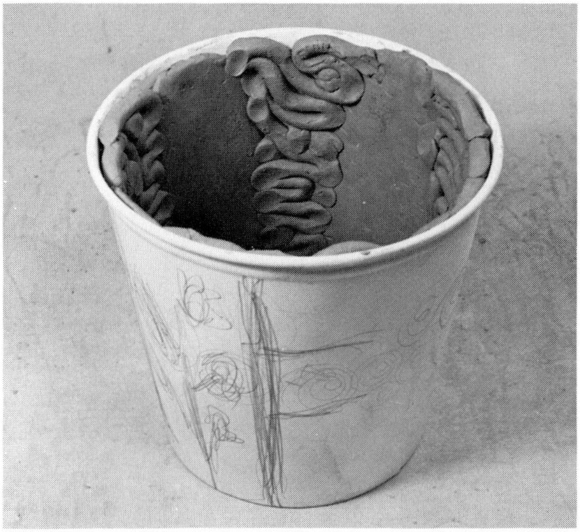

Illus. 40. Place a round slab in the bottom of the container. Place the slabs with the leaves still embedded in them into position on the walls. Fill the areas between the slabs with coils in an irregular design. Knead the slabs and coils together, taking care not to press too hard against the walls.

Illus. 41. When the clay is leather-hard (overnight), turn the container upside-down and lift it off the clay. Pull the embedded leaves out of the clay. (If any small bits remain, they will burn out in the bisque firing.) Then, cut a hole for the lamp stem.

Illus. 42. The lamp base ready to be fired and glazed. See the finished base in color Illus. E on page 18 and Illus. G on page 19.

Illus. Q. This large coil pot by Oscar Bucher was high-fired. The subtle coloring was obtained by the use of oxides and light glazes.

Make a Plaster Mould

Illus. 43. Here are the materials you will need to make your plaster mould.

The best kind of mould for ceramics is a plaster mould because it requires no parting agent, and it will last indefinitely. If you follow the method we describe here, you will be able to make a plaster mould over any simple shape.

To begin, you will need a glass bowl, a ball of clay, a piece of linoleum (or a piece of tin or flexible smooth cardboard), light motor oil, and a smooth hard surface, such as Formica or the smooth side of a piece of Masonite (pressed board).

First place a ball of clay in the glass bowl, making sure the clay is higher than the rim of the bowl. Turn the bowl upside-down on the Formica or Masonite surface. The clay will then hold the bowl in place. Smear a little clay round the edges of the bowl where it meets the hard surface in order to eliminate the undercut of the round edge of the bowl. Remove the excess clay with a piece of cardboard or a modelling tool (Illus. 44).

Lightly oil the entire surface of the glass bowl and an area of about $1\frac{1}{2}$ inches round the bowl on the Masonite surface with light motor oil (Illus. 45). The oil will act as a separator for the plaster.

Illus. 44. Remove the extra clay round the bowl with a tool or piece of cardboard.

Illus. 45. You must oil the bowl so that the plaster will not stick to it.

Roll, tube fashion, a piece of linoleum into a fence shape and place it round the bowl. Fasten the overlapping ends of the linoleum with clothes-pins as shown in Illus. 46. (If using cardboard, oil the inside of the cardboard.) *Leave a space of about 1 inch between the linoleum and the bowl.* This space determines the wall thickness of the plaster mould. Place large rolls of clay at the base of the linoleum fence to keep it from moving and to seal it tight (Illus. 46). To prevent the fence from opening, wind string round it several times and tie.

Mixing the Plaster

A mixture of ordinary casting plaster and water is used to make the plaster mould. The proper

Illus. 46. Large rolls of clay at the base of the fence will keep it from moving.

Illus. 47. As soon as the plaster and water are mixed, pour the mixture into the fenced area, completely covering the bowl. The level of the poured plaster should be approximately 1 to $1\frac{1}{2}$ inches higher than the bowl.

ratio is 50 per cent plaster to 50 per cent water by *weight.*

The most accurate way to measure is to weigh the plaster and water on a scale. However, you can measure the amount of plaster and water by using a plastic bucket of approximately the same size as the linoleum fence. Estimate the total volume of mixture you will need. Sixty-five to 70 per cent of this volume should be water. Add handfuls of plaster to this amount of water, but do not mix yet. Keep adding plaster until about $\frac{1}{8}$ to $\frac{1}{16}$ inch film of water is left. Then mix the plaster thoroughly for $1\frac{1}{2}$ minutes. Whatever way you do it, be sure to *add the plaster to the water.*

Illus. 48. Most plasters set in 20 to 30 minutes. When set, remove the clay and fence. Turn the plaster mould over, remove the bowl, and trim all sharp edges of the mould with a knife. Place the mould in a kitchen oven set at 100° F. (38° C.) for about 48 hours or sun-dry for approximately four days.

Illus. R. A glossy, contrasting glaze adds richness to this filigree coil bowl.

Illus. S. Small wads of clay with a pressed design have been placed in an overlapping fashion to make this charming small pot. The design was enhanced by applying a stain of color which was rubbed off the high areas. A clear glaze was then applied.

Illus. T. The charm of this hanging lamp is due to the casual and irregular design of coils and slotted slabs which have been placed in vertical bands to conform to the vertical shape of the lamp. A cardboard cereal container was used as a mould. (See page 46.)

Illus. U. A light and intricate pattern of coils and textured "buttons" gives this unique mirror frame its distinctive appeal. A dinner plate was used as a mould. (See page 44.)

Illus. 49. Prepare your clay coils and place them inside the plaster mould in an irregular design. The more irregular the placement, the more character the bowl will have. Add small wads or "buttons" of clay for design interest.

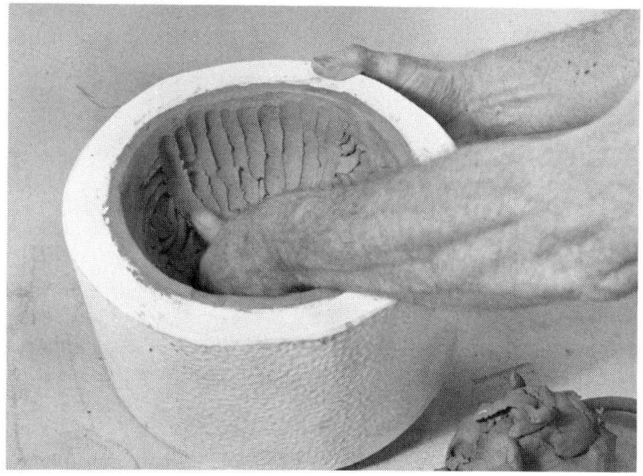

Illus. 50. Knead the coils and "buttons" together using an up-and-down motion, being sure not to press too hard. Add a reinforcing coil on top. You can use a wet sponge to make the inside surface smoother.

Make a Coil Bowl Inside a Plaster Mould

Illus. 51. When the clay is leather-hard (overnight), place a piece of cardboard over the top of the mould, turn the mould upside-down, and lift it off the clay.

Illus. 52. See the finished bowl in color Illus. K on page 22. By using the same mould, you can make variations such as this filigree effect, obtained by leaving open spaces between the coils.

Glazing with Slate and Concrete Sealer

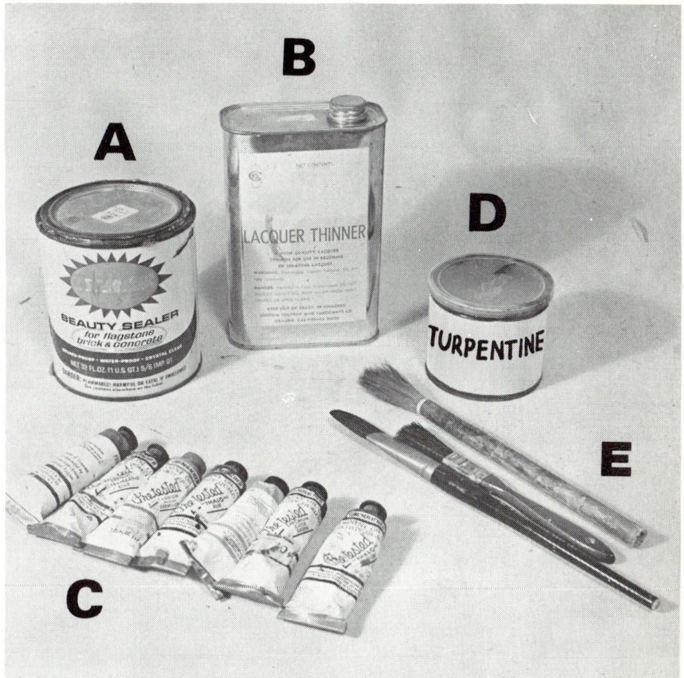

Illus. 53. Materials you need for glazing with slate and concrete sealer: A. slate and concrete sealer; B. lacquer thinner; C. tube oil colors; D. turpentine; E. paintbrushes.

Bisque Firing

Place bone-dry green (unfired) ware in a gas or electric ceramic kiln. You can stack pieces on top of each other—since they are not glazed, they will not stick in the firing. Start the kiln on LOW, leaving the door slightly open for the moisture to escape. In about an hour, turn the heat to MEDIUM, still leaving the door slightly open. In another hour, close the door. In another hour, turn the heat on HIGH. Most bisque firing requires 4 to 8 hours to reach maturing temperature, depending on the size of the kiln and the amount of ware in the kiln.

When the temperature reaches Cone 06, or 1840° F. (1004.4°C.), turn the kiln off. The kiln should be allowed to cool gradually for approximately 8 to 12 hours before the ware is removed.

Glazing

After your coil project has been bisque-fired, you will probably want to glaze it for color and waterproofing.

Ceramic glazes that require a second firing may, of course, be used. However, we would like to introduce you to a different type of glaze that we have used on all of the step-by-step projects illustrated in this book. It is simple enough for everyone to apply, and it is *completely waterproof* and *does not require a firing*.

Illus. 54. Glazing a small pot with colored sealer.

This glaze is basically a *slate and concrete sealer* (Illus. 53A), a clear liquid available at any hardware or paint dealer's. It is inexpensive, odorless, harmless, and dries quickly. However, although the sealer is waterpoof and, in application, non-toxic, it is not hard enough for food and drink containers. For that, you will have to use a fired ceramic glaze.

Be sure that your sealer has an *acrylic base* and not a wax base. (The authors use Trewax Beauty Sealer, since some of the other acrylic sealers are very thin and require a coating of clear acrylic spray for sheen.)

You can use the sealer directly from the can as a clear glaze or you can make any color glaze desired by adding a small amount of tube oil paint (Illus. 53C) to the clear sealer.

Then, simply paint the clear or colored sealer on the fired piece with an ordinary inexpensive paint-brush. When the sealer has dried (one to two hours), your ceramic piece will be covered with a permanent glaze that is completely waterproof.

When glazing the inside of a pot that is to hold water, apply the sealer liberally to make sure all areas are well covered. To ensure maximum waterproofing, it is a good idea to apply two liberal coats of sealer. If the sealer is applied liberally and full strength, a shiny glaze will result. If a matt glaze is desired, thin the sealer with lacquer thinner (Illus. 53B). Use lacquer thinner also as a solvent for cleaning your glaze brushes.

Many color variations can be achieved by the texture of the ceramic piece itself, since the colored glaze will sink into any lines or crevices, making those areas darker.

Try using more than one color glaze on your piece. For example, let a second color glaze drip partially down over the first color glaze, or, apply one color glaze over another (such as green over blue) for a deeper, richer color.

Another easy and effective way to color your fired ceramic piece is to paint it with a thin wash of oil color and turpentine (Illus. 53D) and then, when the wash has dried (a matter of minutes), glaze the entire piece with clear or tinted glaze. By using the oil color-turpentine wash, it is possible to shade one color into another, as in a painting.

While children especially will enjoy glazing their ceramic pieces with either sealer-glaze method because it is so easy, and results can be seen immediately, everyone, whether beginner or experienced, now can create a permanent, water-proof glaze of any desired color without a glaze firing by using slate and concrete sealer.

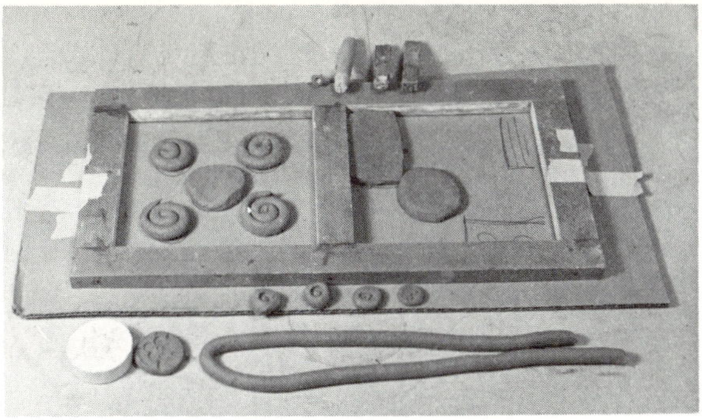

Illus. 55. Make a wooden frame about ½ inch deep for one or two tiles, and tape it to a flat piece of cardboard. You can use the frame over again to make as many tiles as needed. Draw a rough design on the cardboard and texture a number of coils, "buttons," and slabs.

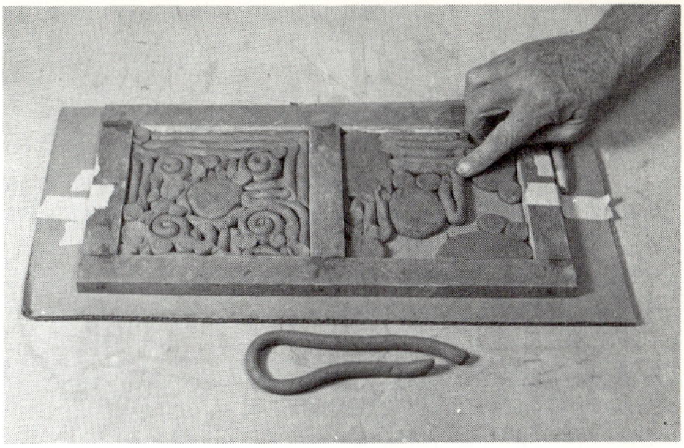

Illus. 56. Place the textured coils, "buttons," and small slabs inside the wooden frame, not following the design too closely. You will have greater success if you stay loose and improvise as you go along.

Make a Tile Wall Plaque

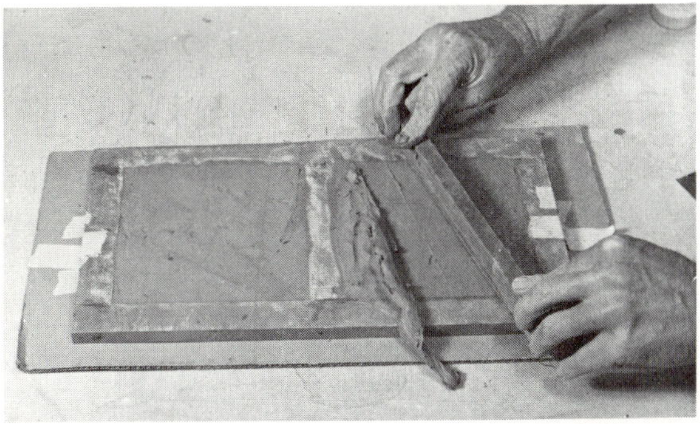

Illus. 57. After you have kneaded the coils (without too much down pressure), build up the clay to the top of the wooden frame by "plastering" with soft clay. Remove the excess clay with the edge of a straight stick. This will level the tiles to an even thickness.

Illus. 58. When the clay is leather-hard and has shrunk away from the frame, remove the frame. To prevent warping, place a board on the tiles until dry. If you are making a six-tile plaque such as in color Illus. F on page 18, speed up the drying by placing the clay still in the frame in the sun or under a heat lamp for about 30 minutes.

Illus. 59. Start a free-form pot in the usual way (see page 9). After reaching a height of 2 or 3 inches, knead the coils together on the inside and the outside, too. Continue this procedure until you reach the desired height.

Make a Free-Form Pot with a Smooth Finish

Illus. 60. When you reach the desired height, smooth and shape the pot with a wooden modelling tool.

Illus. 61. To make a narrowing top, lay a few more coils slightly inside of each other as shown. Then knead these coils together on the inside and outside.

Illus. 62. To complete the shape, push the walls outward with a wet paintbrush. Then smooth the entire surface with a wooden modelling tool and a wet sponge.

Illus. 63. For extra interest, attach with slip thin slabs of clay with an incised leaf design and stamped "buttons" (see color Illus. 0 on page 26). With this method, you can easily make smooth pots without having to use a potter's wheel.

Illus. 64. Place your slabs, texture-side out, in a fish-scale (overlapping) fashion at irregular intervals, both horizontally and vertically.

Make a Coil and Slab Pot Inside a Wooden Planter

First dust an *unpainted* and *unsized* wooden planter with powdered clay, and then line the bottom with paper. Then prepare a quantity of coils and roll out a number of different-sized slabs (see page 28) and texture the slabs with your texture tools or corrugated cardboard.

If you are handy with a hammer and saw, you can make your own wooden moulds of all shapes and sizes. Wooden moulds are perfect for this kind of coil work, and they last indefinitely.

Illus. 65. Then fill in the open areas between the slabs with coils in an irregular pattern to make the design more interesting. As always, take care not to press the coils and slabs too firmly against the wooden walls or the design might be lost.

Illus. 66. When the slabs and coils are all in place, fill in any open spaces with small wads or "buttons" of clay. Then knead the entire surface as usual. Add a reinforcing coil and smooth the top with a wet sponge.

Illus. 67. As soon as the clay is leather-hard, place a piece of cardboard on top, turn the wooden planter upside-down and lift it off the clay. Cut a drainage hole in the bottom with a knife. Turn the pot right-side up and smooth the top with a wet sponge. See a similar pot in color Illus. L on page 23.

Illus. 68. Place a dinner plate upside-down on a paper towel and draw the outline of the plate. The paper is absolutely necessary to keep the clay from sticking to the nonporous, glazed plate. You will use a small, round mirror in the frame.

Make a Mirror Frame from a Dinner Plate

Illus. 69. Turn the plate right-side up. Cut the paper towel and tape it to the edges of the plate. Place clay coils and "buttons" irregularly on the paper surface. The underside of the "buttons" shown here in place were textured with a bolt.

Illus. 70. When all the coils and buttons are in place, knead them together. Here, the edges of the plate have been left irregular for interest. Add a reinforcing coil as shown to prevent the irregular ends of the coils from cracking. Then draw the shape of the mirror on the kneaded surface.

Illus. 71. For hanging purposes, use slip to attach small wads of clay with holes that will accommodate picture wire. Place a coil round the drawn shape of the mirror so that the mirror will not slip when put in place. Then cut out and remove the mirror shape, but cut it a little smaller ($\frac{1}{8}$ inch) than the outline.

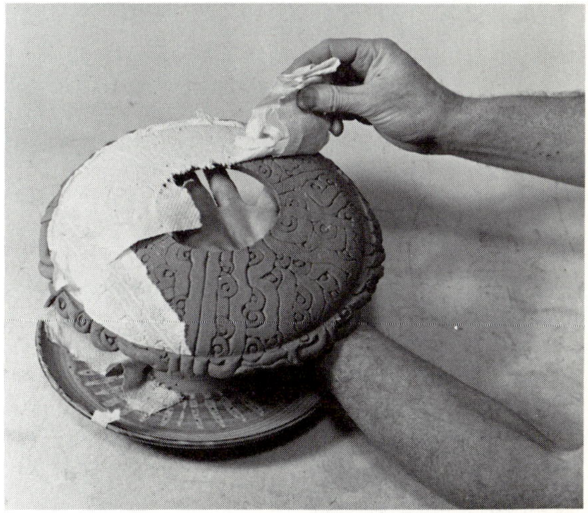

Illus. 72. When the clay is leather-hard, lift it off the plate and carefully remove the paper towel.

Illus. 73. When the clay frame has been fired and glazed, glue the mirror in place with white glue or epoxy cement. (See the finished frame in color Illus. U on page 35.) You can also use an unwaxed paper plate to make the frame, thus dispensing with the paper-towel liner. Also, you may prefer to use the concave side of a plate to make your frame.

Make a Hanging Lamp in a Cardboard Container

Illus. 74. Make a rough sketch as a design guide for your lamp. Dust inside of cardboard container with powdered clay and line the bottom with paper. Then make a circle of clay coils for the bottom, and place in the container.

Illus. 75. Next, cut coils of clay into short lengths (about 2 inches). Place these vertically in the container to form the first band of clay.

Illus. 76. Roll out narrow slabs of clay (page 28) and slot them with a wire modelling tool, following your drawn design. The slots will allow light to show through the finished lamp.

Illus. 77. Place a slotted strip in the container to form the second band. Knead the bottom edge of the strip to the band of vertical coils.

Illus. 78. Add another band of vertical coils and another slotted slab. Knead these together in the same way as the first two bands. Add a final band of vertical coils, knead, and add a reinforcing coil. When kneading, be careful not to knead across the open slots.

Illus. 79. When leather-hard, turn the cardboard container upside-down and lift it off the clay lamp. If the container does not slip off easily, let the clay dry longer (another 4 to 6 hours depending on weather and room temperature). Remove the paper circle and drill or cut a hole in the middle for the lamp cord. See the finished lamp in color Illus. T on page 35.

INDEX